MAJOR DISASTERS

INDIAN OCEAN TSUNAMI

BY ANNETTE M. CLAYTON

WWW.APEXEDITIONS.COM

Apex is distributed by North Star Editions:
sales@northstareditions.com | 888-417-0195

Produced for Apex by Red Line Editorial.

Photographs ©: zz/Stephen Trupp/STAR MAX/IPx/AP Images, cover, 4–5; Shutterstock Images, 1, 13, 16–17, 21, 22–23, 25; David Rydevik/Wikimedia Commons, 6; iStockphoto, 7, 12, 29; Gemunu Amarasinghe/AP Images, 8–9; NASA, 10–11; Bedu Saini/Serambi Indonesia/AP Images, 14–15; Dudi Anung/AP Images, 18; Deddeda Stemler/CP/AP Images, 20; Andy Eames/AP Images, 24; Irwin Fedriansyah/AP Images, 26–27

Library of Congress Control Number: 2023910182

ISBN
978-1-63738-759-7 (hardcover)
978-1-63738-802-0 (paperback)
978-1-63738-885-3 (ebook pdf)
978-1-63738-845-7 (hosted ebook)

Printed in the United States of America
Mankato, MN
012024

NOTE TO PARENTS AND EDUCATORS

Apex books are designed to build literacy skills in striving readers. Exciting, high-interest content attracts and holds readers' attention. The text is carefully leveled to allow students to achieve success quickly. Additional features, such as bolded glossary words for difficult terms, help build comprehension.

TABLE OF CONTENTS

CHAPTER 1

GIANT WAVE

It's a beautiful day in Thailand. Crowds relax on the beach. Suddenly, the water **recedes**. Fish flop on the wet sand. People stare at the strange sight.

Thailand was one of 14 countries hit by a tsunami in December 2004.

Next, a giant wave rushes toward the shore. It's a tsunami! Everyone runs, but the wave is very fast. It sweeps some of them away.

Some people near the coast took photos as the tsunami's first wave hit.

To stay safe in a tsunami, people must get to higher ground.

KID HERO

Ten-year-old Tilly Smith was on vacation in Thailand. She had studied tsunamis in school. When the water receded, she knew danger was coming. She and her parents warned and saved 100 people.

Several more giant waves follow. The deep water continues moving inland. It rips up trees and topples buildings.

FAST FACT

Some waves were more than 160 feet (49 m) tall.

A tsunami's fast-moving water sweeps away everything in its path.

CHAPTER 2

HOW IT HAPPENED

The tsunami took place on December 26, 2004. An underwater **earthquake** caused it. The earthquake shook the seafloor. It released huge amounts of energy.

The earthquake happened in the Indian Ocean near the west coast of Sumatra.

A tsunami often has several waves. The first ones may not be the biggest.

Waves formed at the top of the water. They rippled out in many directions. About 20 minutes later, the first wave crashed into Indonesia.

WAVE SCIENCE

Tsunami waves start out in the ocean. At first, they are short and far apart. As they near the coast, they slow down. They grow taller and closer together.

The first wave hit Banda Aceh, Indonesia. This area was very badly damaged.

A tsunami's rushin
water can drag
people along with

Giant waves also hit Thailand, India, Sri Lanka, and Maldives. Deep water flooded the coasts. Then it rushed back to the ocean. It pulled people and **debris** with it.

FAST FACT

Some tsunami waves reached all the way to East Africa. They caused floods there, too.

CHAPTER 3

DAMAGE

The event was the deadliest tsunami in history. Nearly 230,000 people were killed or missing. Some drowned. Others were hit by debris the water carried.

Indonesia had the worst damage. More than 167,500 people there died or went missing.

The floods and debris damaged large areas of land. Entire towns were smashed into **rubble**. Chunks of houses and trees floated down roads.

FAST FACT

Many hospitals were damaged. Wounded people could not get help.

In some towns, few or no buildings were left standing.

Waste and debris made the water dangerous. Nearly two million people had to leave their homes. Many farm fields were damaged, too.

Water left by a tsunami often becomes filled with trash and germs.

Many people in Indonesia grow rice. Covering rice fields in fresh water helps the plants grow.

SALTY WATER

Salty ocean water leaked into farmlands. It killed crops. As a result, many people ran out of food. Salt water also got into pipes and wells. Drinking water became unsafe.

CHAPTER 4

AFTER THE TSUNAMI

Many countries sent aid. They helped people have clean water and safe places to stay. People also worked to rebuild damaged areas.

To reach flooded areas, groups often sent aid on planes and helicopters.

Some aid focused on the **environment**. The tsunami **eroded** coasts. It also knocked down many mangrove trees. People worked to **restore** these natural areas.

Lots of land, roads, and bridges near coasts washed away in the tsunami.

People plant new mangrove trees near coasts. The trees help block tsunami waves.

MANGROVES

Mangrove trees have long, tangled roots. They **filter** water as it flows into the ocean. They also help protect against tsunamis. Their thick roots break up waves.

A worker helps test a tsunami warning system in Indonesia.

FAST FACT

Some areas have safety drills. People practice what to do if a tsunami hits.

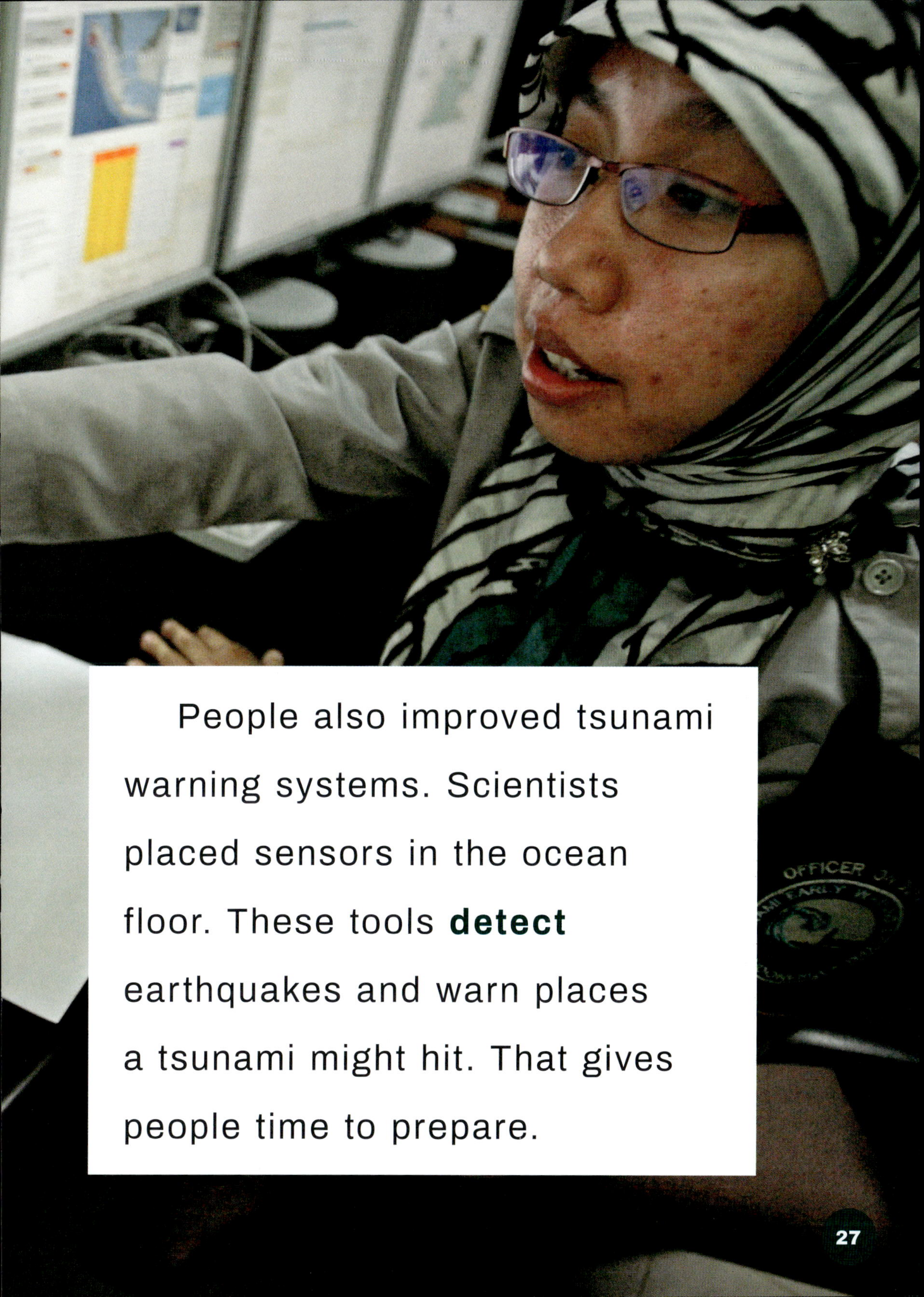

People also improved tsunami warning systems. Scientists placed sensors in the ocean floor. These tools **detect** earthquakes and warn places a tsunami might hit. That gives people time to prepare.

COMPREHENSION QUESTIONS

Write your answers on a separate piece of paper.

1. Write a few sentences describing how and why the tsunami formed.

2. Would you like to work as a scientist who studies tsunamis? Why or why not?

3. How tall did the tsunami's tallest waves get?

- **A.** less than 26 feet (8 m)
- **B.** more than 160 feet (49 m)
- **C.** more than 500 feet (152 m)

4. What would likely happen if a coast had fewer mangrove trees?

- **A.** Tsunamis would cause more damage there.
- **B.** Tsunamis would cause less damage there.
- **C.** Tsunamis could no longer happen there.

5. What does **crops** mean in this book?

Salty ocean water leaked into farmlands. It killed ***crops****. As a result, many people ran out of food.*

A. plants people grow for food
B. places where people build their homes
C. animals that people ride

6. What does **aid** mean in this book?

Many countries sent ***aid****. They helped people have clean water and safe places to stay.*

A. money kept in a bank
B. money paid to a country's government
C. help given to people in need

Answer key on page 32.

GLOSSARY

debris

Pieces of something that broke or fell apart.

detect

To sense or find something.

earthquake

An event where part of Earth shakes or trembles.

environment

The natural surroundings of living things.

eroded

Wore away.

filter

To remove harmful things from something.

recedes

Moves backward.

restore

To help something become strong or healthy again.

rubble

Pieces of buildings that have fallen apart.

TO LEARN MORE

BOOKS

Adamson, Thomas K. *The Indian Ocean Tsunami*. Minneapolis: Bellwether Media, 2022.

London, Martha. *Tsunamis*. Minneapolis: Abdo Publishing, 2020.

Rathburn, Betsy. *Tsunamis*. Minneapolis: Bellwether Media, 2019.

ONLINE RESOURCES

Visit **www.apexeditions.com** to find links and resources related to this title.

ABOUT THE AUTHOR

Annette M. Clayton is a writer living in Maryland with her husband and twin daughters. She enjoys hiking the Appalachian Trail, traveling, and writing books that teach kids about the fascinating world around them.

INDEX

ANSWER KEY:
1. Answers will vary; 2. Answers will vary; 3. B; 4. A; 5. A; 6. C